SOLVING THE MYSTERIES OF DIGESTION

In this Health Guide, Jeffrey Bland makes very clear the anatomy of the digestive system, so often confused by technical terminology. The journey of the food we eat from digestion to absorption to elimination follows its logical, sensible map, easy to follow and become familiar with. Tracing the phases of digestion—beginning with the cephalic phase which is the start of digestion before you take the first bite—the work of digestive enzymes, the problems caused by fasting or those involved in an ailing pancreas is made a fascinating expedition, full of new and helpful information.

ABOUT THE AUTHOR

Jeffrey Bland, Ph.D. is CEO of HealthComm International, Inc., a leading global health communications firm, located in Gig Harbor, Washington. Before founding HealthComm, Dr. Bland was for 13 years a professor of chemistry at the University of Puget Sound in Tacoma, Washington. He is a former Senior Research Scientist at the Linus Pauling Institute of Science and Medicine and former director of a large medical laboratory in Washington State. Dr. Bland lectures widely on health and nutrition and has written and edited numerous books and articles for the scientific and popular press, including several works in the Good Health Guide series and the second of Keats Publishing's annual survey series, *A Year in Nutritional Medicine.*

Digestive Enzymes

20 million people suffer
from digestive disorders.
Are you one of them?

Jeffrey Bland, Ph.D.

Keats Publishing, Inc. New Canaan, Connecticut

DIGESTIVE ENZYMES

Copyright © 1993 by Jeffrey Bland

ISBN: 0-87983-331-9

Printed in the United States of America

Keats Good Health Guides™ are published by
Keats Publishing Inc.
27 Pine Street (Box 876)
New Canaan, Connecticut 06840-0876

Contents

INTRODUCTION

It has often been said that you are not what you eat, but rather what you absorb of your diet. You can eat the best of foods and take the most well-balanced of nutritional supplements but if you do not digest and absorb these substances you will suffer from malnutrition, and enjoy less than optimal health. Diminished immunity, delayed wound healing, allergies, muscle wasting and poor bowel function may all result from a maldigestion and/or malabsorption problem. We commonly think of digestive disorders as resulting in stomach or abdominal pain, but symptoms such as mood swings and skin problems may also be a direct consequence of the maldigestion and malabsorption of specific nutrients.

It is suggested that as many as twenty million Americans suffer from various digestive disorders which impair their nutrition.[1] For these people, inspection of their diets, lifestyle, and the need for enzyme supplementation may mean the difference between a healthy, productive life and a life marked by ill health and low vitality. In order to understand this problem, we need to know something about the "plumbing system" of the body and how it is used to digest food and promote the absorption of nutrients. We will also learn how optimal health is related not only to eating high-quality foods but also to proper digestion, absorption and distribution of nutrients and excretion of the final waste products.

The first step is the understanding of the parts of your digestive system and how they operate as a team.

ANATOMY OF THE DIGESTIVE SYSTEM

The function of the gastrointestinal system, which includes the mouth, esophagus, stomach, small and large intestines, salivary glands, portions of the liver, pancreas and gallbladder, is to break down the big materials in your diet (proteins, carbohydrates and fats) into the small, absorbable components, amino acids, sugars and free fatty acids. These small substances can then be taken into the blood across the intestinal barrier, where they promote nutrition of the whole body. The intestinal system is somewhat like a household plumbing system; it is a barrier that defends the rest of our bodies from exposure to toxins and other potentially adverse materials, but can also prevent the proper absorption of nutrients if it is not functioning correctly.

Contrary to popular belief, the intestinal system is not actually a major avenue for eliminating waste from the body. In people who have proper digestion and absorption only small amounts of excretory products are normally eliminated in the feces, and most of the fecal material is made up of dead bacteria from the intestinal tract, the small amount of blood, bile, and other materials that have been broken down in the intestines and indigestible dietary fiber. The major routes of excretion of waste products from the body are through the skin, the lungs and the kidneys. A properly functioning digestive tract should not produce fecal material with a lot of undigested, unabsorbed food. The digestive action of the intestinal tract, which is about twenty-five feet long, is controlled by the secretion of a number of hormones and substances called enzymes which participate in breaking down the food materials into their absorbable components as well as the presence of agents which help emulsify fats, putting them into forms which can be readily absorbed into the blood.[2]

The intestinal tract is home to millions of bacteria. They are normally harmless, and even beneficial in that they may produce some vitamins which you can absorb and utilize. These bacteria, however, are affected by the state of your intestinal tract, and if it is less than healthy they can go from being friends to becoming foes and produce disease. Overgrowth in the intestines of certain disease-producing bacteria is a well-known cause of digestive disorders and may also bring about serious complications, which will be discussed later in this Health Guide.

The absorption of nutrients across the intestinal boundary depends upon the health of the mucous lining of the small intestine. This lining is deeply folded and covered with millions of tiny projections called villi. It is through the cells lining the villi that absorption of nutrients occurs. Different regions along the intestinal tract absorb specific nutrients, so that loss of integrity of the villi in one area (for instance, in the case of inflammation) may cause malabsorption of only one nutrient, while absorption of other nutrients elsewhere in the intestinal tract (where there is no inflammation present) will not be affected.[3]

Some patients thus experience selective malabsorption and demonstrate a specific nutrient deficiency of one or more types, rather than generalized malnutrition.

One other feature of the intestinal tract is the last meter and a half—about five feet—of the large intestine, called the colon. The colon is extremely important in nutrient utilization and can even contribute to the body's stores of B vitamins. It does so by way of what are called symbiotic bacteria, living in harmony with the body and producing their own vitamins, which are absorbed into the blood and used as a source of nutrition by the host.

We have a very subtle and important relationship with the bacteria that live in our colons. It has been estimated that there are more bacteria living in the colon than there are cells that make up the whole body. When these bacteria are living in proper symbiosis (meaning they contribute to us and we provide an environment for them), we have a healthy intestinal tract, efficient elimination of waste materials and good nutrient absorption. If, however, the intestinal bacteria are in a state of disarray, they may go from being symbiotic to becoming parasitic. In this case, a toxic reaction can develop in the colon, with symptoms of intestinal discomfort, gas formation, constipation and poor nutrient utilization.

Toxic colon disorders are associated with higher levels of various anaerobic bacteria (those that do not require oxygen). These organisms, such as *Clostridium difficile* and *Clostridium perfringens*, are known to manufacture various by-products from undigested food, which may be absorbed into the bloodstream and lead to toxic-like reactions.

The integrity of the bacteria in the colon is controlled by the acidity of the mucosa of the colon, as well as by fiber in the diet. On occasion, reinoculation of the colon with an oral preparation of *Lactobacillus acidophilus* or *bulgaricus* will help to reestablish colonic bacterial action, which is important for the maintenance of bowel regularity, for dewatering of the stool for the right consistency and for bile salt reabsorption.

Although the colon represents only a small percentage of the total length of the intestinal tract, it is very important in the maintenance of nutrient balance and gastrointestinal function. If a considerable amount of undigested and unabsorbed food arrives in the colon with the food mass, the bacteria can multiply rapidly and produce potential toxic metabolites directly in the colon, some of which may even be carcinogenic (cancer-producing). This is why it's extremely important to have sufficient digestive enzyme secretion, so that protein, carbohydrate, and fat can be broken down and assimilated in the small intestine before they arrive in the colon.

As mentioned earlier, fecal material is almost entirely made up of water, dead bacterial cells, breakdown products of hemoglobin, bile salts and undigested food fiber. It should not be composed of unassimilated food. If the stool is fatty in composition or contains undigested food fibriles, enzyme deficiencies are a possibility. Hence, examination of fecal material can be very useful in detecting malfunctions of certain aspects of intestinal physiology, even in the small intestines.

THE IMPORTANCE OF DIGESTIVE ENZYMES

The breaking down of food materials such as protein, carbohydrate and fat into smaller absorbable units is dependent upon the secretion in the digestive tract of a class of proteins called enzymes, which are responsible for your ability to absorb nutrition from your foods. There are specific digestive enzymes that break down protein called proteases; carbohydrate, called amylases; and fat, called lipases. The first secretion of a digestive enzyme when you eat, occurs as you are chewing, when salivary amylase, an enzyme that can break down starch, is secreted in the mouth. You do not start to digest protein or fat in your mouth, but that is where starch digestion is initiated. One of the important reasons for thoroughly chewing your food before swallowing is to get a good start with starch digestion.

Cephalic phase of digestion: It's important to note that digestion of your food begins before you even take the first bite. As you think about a meal, you are setting the stage for its digestion when you actually eat it. This is called the cephalic (from the Greek word for "head") phase of digestion. If you are thinking good thoughts and prepared for a fine meal, you are already secreting digestive enzymes and hydrochloric acid in your stomach, which will promote digestion of those nutrients once they arrive in the stomach. If, however, you are uptight and anxious before your meal, or are under severe stress, you may not be able to secrete enough acid or enzymes, and this negative cephalic phase can result in maldigestion. This explains why a lot of people suffer from digestive upsets and intestinal problems during periods of extreme stress.[4]

Gastric phase of digestion: After the food has entered the mouth and initiated the digestive process, the body moves into the gastric phase of digestion. In this phase the food travels to the stomach and specific hormones are released which initiate the enzyme breakdown of foods. One of the most important hormones is gastrin, which is released from the cells in the back of the stomach; this release is initiated by dietary protein. Meals that are too low in protein do not stimulate the secretion of gastrin, and this may result in poor digestion. Gastrin stimulates cells within the stomach called the parietal cells to secrete hydrochloric acid, which causes the food in the stomach, now starting to be digested, to become acidic, which is essential for its full digestion. A protease enzyme called pepsin is also released and begins the breakdown of protein.

Most of the digestive enzymes such as pepsin are secreted initially in inactive forms called zymogens or proenzymes. This prevents the body from digesting itself. Names of zymogens always end in the suffix -ogen—e.g., pepsinogen, trypsinogen and chymotrypsinogen. Conversion of these enzymes to their active forms requires adequate levels of other converting enzymes, called coenzymes, which are affected by the levels of zinc and manganese in the body. Deficiencies of these trace minerals can therefore result in digestive problems.

When food enters the stomach, the acidity of the gastric environment may be quite high, sometimes so high that many enzymes present in the food are denatured and become ineffective. This is one of the reasons that many vegetable enzymes, such as papain and bromelain, have been found not to be as useful in promoting digestion of food as animal-derived enzymes such as chymotrypsin or trypsin. Amino acids which make up the protein of foods—for example, phenylalanine and tryptophan—have been found to stimulate the release of digestive enzymes such as pepsin, which is one reason why dietary protein is important to nutrient utilization.[5] After the food has remained in the stomach for a period of time, it is pushed by muscle contraction into the small intestine, the first region of which is called the duodenum.

Intestinal phase of digestion: Once the food has been moved into the duodenum, the intestinal phase of digestion is initiated.

This is when the major breakdown of protein, carbohydrate and fat to absorbable units occurs. The partially digested food mass coming from the stomach must be acidic in order to evoke the output of the digestive enzymes which are responsible for the major breakdown of the food materials. These enzymes are the proteases chymotrypsin and trypsin, which break down protein; amylases and saccharidases, which break down various types of dietary fats, including triglycerides and phospholipids. They are secreted by the same organ, the pancreas, and released into the duodenum. (This takes place in what is called the *exocrine* portion of the pancreas; the *endocrine* portion secretes the hormones such as insulin and glucagon which help control blood sugar.) This secretion of enzymes by the pancreas is simulated by two hormones called secretin and cholecystokinin, which themselves are secreted by the lining of the duodenum. The release of these hormones is triggered by the acidity of the intestinal contents. If the stomach was not able to acidify the food adequately, due to parietal cell insufficiency, it might not stimulate release of sufficient digestive enzymes, with resulting maldigestion or malabsorption.[6]

The pancreas secretes these enzymes in a matrix of sodium bicarbonate, which acts to alkalize the very acidic food material. If there is not enough bicarbonate released, the enzymes cannot work effectively in the duodenum to break down foods. We thus have the paradoxical situation of an *under*acid stomach producing an *over*acid small intestine, which renders the enzymes less able to make nutrients absorbable.

It is clear that stomach acid secretion is essential for effective nutrient absorption; yet, as people get older, they often lose the ability to secrete this acid, and consequently many older individuals have malabsorption problems. In these cases stomach acid replacement therapy may be necessary, utilizing supplements such as betaine hydrochloride or glutamic acid hydrochloride as acidifying materials.

Effect of fasting on digestion: When an individual fasts, obviously the stomach is no longer loaded with food and the pancreas need not secrete enzymes and bicarbonate. It has been found that over a period of fasting the pancreas may lose its ability to secrete adequate levels of amylase, the starch-reducing enzyme, so that when the individual eats again, if the meal is high in starch, he may be unable to digest it effectively and the meal produces large amounts of gas.[7] Gas formation is primarily the result of undigested and unabsorbed nutrients traveling to the colon, where bacteria ferment these nutrients, producing gas as a byproduct. The most common gases produced ·by bacteria in the intestines are the odorless methane and hydrogen. If there are considerable amounts of odoriferous substances produced by fermentation, this generally indicates that sulfur-rich proteins are being maldigested. These proteins may come from beans, onions, garlic, eggs or meat. Excessive gas or flatulence indicates a possible maldigestion or malabsorption syndrome or the presence in the diet of usually poorly absorbed or metabolized sugars such as raffinose, which is abundant in beans, or lactose, which is abundant in milk. Drs. Grossman and Ivy have found that when an individual fasts, ends a fast, or otherwise changes diet drastically, the pancreas has to adapt to the new conditions by varying its secretion of enzymes.[8]

When an individual habitually eats meals high in protein, the pancreas will secrete up to seven times the normal amount of enzymes capable of breaking down protein; but if his regular diet is high in starch, his pancreas will be up to ten-fold higher in amylase secretion activity. This seems to explain why certain individuals, after a fast or a radical change of types of foods, experience digestive complaints. The pancreas may have adapted to produce enzymes more consistent with one diet than another and the addition of the new dietary material leads to some maldigestion or malabsorption and such symptoms as intestinal discomfort, flatulence or bloating.

The conclusion to be drawn from this is that major diet alteration should occur slowly, not overnight. Even the addition of dietary fiber should occur in small increments, by adding a teaspoonful of supplemental fiber each day rather than starting the first day with three to five tablespoonsful. It is also important to note that there are foods which contain enzyme inhibitors, which may act to prevent the breakdown of fat, protein or carbohydrate. We know that soybeans contain a trypsin inhibitor that impedes the digestion of protein, and that soybeans should therefore be thoroughly cooked to prevent this factor from having a digestion-blocking influence. We also know that certain beans or peas contain amylase inhibitors which can prevent the effective digestion and absorption of starch. This fact gave birth to the use of orally administered starch blockers to cause weight loss; however, as has been recently pointed out in three studies, oral starch blockers were not effective in facilitating weight loss by preventing the starch-derived sugars from being absorbed into the bloodstream. [9,10,11]

Effects of fluids on digestion: It is known that the heaviest secretion of enzymes from the pancreas occurs one to two hours after a meal and that fluids taken with the meal stimulate gastric secretion for optimal acidity in the stomach and promote enzyme and bicarbonate secretion in the small intestine.[12] This contradicts the seemingly logical notion many people have, that fluids with the meal dilute the enzymes and prevent proper digestion, but the evidence seems quite clear that sufficient fluid with the meal is essential for stimulating gastric, pancreatic and intestinal responses. The liquid should be preferably at room temperature, in the amount of one or two glasses, for optimal intestinal function.

It is also known that the acidity of a food can alter its digestive capability. The more acid the foods, such as citrus fruits or acid vegetables, the slower the emptying of the stomach and the longer the digestive phase.[13] This suggests that supplementation with vitamin C may slow the digestive process and prolong the release of nutrients into the bloodstream across the intestinal border. The vitamin C is an acidic substance and may contribute to slowing the passage of the stomach contents into the intestines for digestion. Vitamin C as the ascorbate (sodium, calcium) will not produce this effect.

Pancreatic insufficiency: Although the pancreas has a remarkable ability to secrete enzymes, and a 90 percent loss of the enzyme-secreting pancreatic acinar cells must occur before actual clinical signs of fat malabsorption syndrome (fat in the stool or undigested protein in the fecal material) appear, there is evidence that chronic pancreatic enzyme insufficiency may contribute to a variety of adverse health effects well before this indication of pancreatic disease is seen.[14]

Since adequate enzyme activity is necessary for vitamin B12 absorption, possibly insufficiencies of digestive enzymes are involved in symptoms of vitamin B12 deficiencies such as anemias and peripheral pain in the hands and feet. Clinical signs of pancreatic enzyme insufficiency, which include flatulence after meals, swelling of the abdomen, skin problems, recurring headaches, muscle wasting and even depression, may indicate the need for enzyme replacement therapy.

ENZYME REPLACEMENT THERAPY

Enzyme replacement therapy employs digestive enzymes isolated from bovine or porcine sources, administered orally. The use of these enzyme products from the pancreas of cattle or hogs helps improve the digestive and absorptive sufficiency of individuals who exhibit malabsorption syndrome due to pancreatic insufficiency. These products are generally known as pancreatins. It seems hard to believe that an orally administered enzyme can survive its journey through the highly acidic environment of stomach and into the intestine with enough "live" activity remaining to facilitate the digestion of fat, protein or carbohydrate, as acid is generally known to denature enzymes—that is, render them biologically unuseful. However, research and testing has shown that these preparations do manifest varying amounts of effectiveness.

Dr. David Graham of the Department of Medicine, Baylor College of Medicine, recently examined sixteen commercially available pancreatic enzyme extracts, and found that they were able to improve digestion and assimilation of protein and fat in people with pancreatic insufficiency when taken orally. He also found a very wide range of activity among the various preparations studied.[15] Figure 2 shows the dramatic difference in the potencies of these preparations. The highest-potency preparations were found to have the greatest ability to improve digestive function. These preparations were those manufactured by processes that utilize lower temperatures, no acids or alkalies, and no large amounts of salt. The gland of the animal whose enzymes are to be concentrated and put in tablet form needs to be treated with care, because enzymes are easily denatured by heat. If temperatures exceeding 75 to 90 degrees Fahrenheit are used in the manufacturing process, this will result in considerable denaturation of the enzymes in the concentrate and produce a low-activity digestive aid.

When comparing capsules, tablets, and enterically coated tablets, it was found that the capsules and tablets were generally the most effective in optimizing the digestion of fat, carbohydrate and protein. In the enteric-coated substances, the coating often did not dissolve early enough in the intestines to promote digestion, so that much of the enzyme's power was lost. Vegetable enzymes such as bromelain and papain were found by Dr. Graham not to be very effective in comparison to the animal-derived enzymes. The highest-activity enzyme preparations were those that had the highest NF units of activity, with 5 NF (National Formulary) potency units showing the greatest ability to improve nutrient absorption.

In a recent paper by Dr. Eugene DiMagno of the Mayo Clinic, it was stated that orally administered digestive aids such as pancreatin or 5NF were extremely useful in preventing protein or fat maldigestion/malabsorption in people who had frank diagnosed malabsorption problems—that is, clear indications of pancreatic disease.[16] Dr. DiMagno and his colleagues used twelve to sixteen tablets of pancreatin extract with each meal for such patients, though the requirements for people with chronic pancreatic insufficiency may be one to two tablets with each meal. The use of this type of supplement seems to be very helpful in the management of pancreatic insufficiency-induced symptoms such as irritable bowel problems or flatulence.

These oral pancreatic enzyme aids are generally given along with meals and sometimes have to be used in conjunction with acid replacement therapy. Betaine hydrochloride or glutamic acid hydrochloride can be administered with the oral pancreatic enzyme preparations during the meal. Acid should not be administered on an empty stomach, nor should pancreatic enzymes, for fear of irritating the lining of the stomach.

An interesting use of this therapy is in the management of childhood asthma and eczema. Dr. Bray found many years ago in studying juvenile asthmatics and eczemic children that the use of acid-replacement therapy can alleviate the symptoms of these conditions, presumably by improving digestion and assimilation of various nutrients.[17]

Dr. Jonathan Wright of Kent, Washington, has recently used this therapy with great success in the alleviation of symptoms of juvenile asthma. The low stomach-acid secretory ability of the individual may be one of the major problems with this condition, and the acid may have to be administered in conjunction with pancreatic enzyme replacement therapy.

PANCREATIC INSUFFICIENCY AND ITS RELATIONSHIP TO FOOD ALLERGY

Recently Dr. W. A. Hemmings suggested that absorption into the blood of partially digested dietary protein may induce a food-allergic response which occurs either directly at the intestinal barrier or possibly more systemically, resulting in skin problems, headaches, recurrent infection or psychiatric disturbances.[18] He terms these substances incomplete protein breakdown products (IPBs). It is suggested in most medical textbooks that the intestinal lining plays a protective role for the body in preventing the absorption of large dietary substances such as protein.

This barrier has recently been found to be incomplete, since various large substances and particles derived from food have been shown to cross the intestinal boundary. This passage of protein or other incompletely digested food components into the bloodstream may occasion immune reactions which we call allergy. These substances, once they enter the blood, may act as antigens—allergy-producing substances—and there are now two routes acknowledged by which these agents might travel through the intestinal lining. The first route is via the covering of the lymphatic tissues in the intestinal tract. The second is by what is called M-cell vesicle formation and the transport of large materials directly into the blood by a microenvagination process. Doctors Cornell and Iselbacher have found that even very large proteins which may be consumed in the diet, such as the enzyme from horseradish root called peroxidase, can be partially taken up into the blood intact after a meal, suggesting that a small amount of dietary protein may not be completely broken down in the intestines before absorption into the blood.[19]

This is very important clinically in that it indicates that maldigestion of dietary protein may lead to the symptoms of allergy or food hypersensitivity. This accounts for the beneficial effect of administration of digestive enzymes such as pancreatin to people suffering from wheat (gluten) sensitivity, who many times have their symptoms of allergy eliminated by the addition of the digestive aid.

Recently it has been found that certain dietary protein molecules, when incompletely digested after the meal, may cause a reduction in the capacity of the immune system to respond appropriately, leading to long-term allergic reactions by producing a state of low zone tolerance.[20] Even microamounts of undigested protein, if absorbed in the blood, may have profound allergy-producing effects. This process of incomplete dietary protein breakdown and subsequent absorption may be the means by which allergy is initiated in infants breast-fed by mothers who are eating allergy-producing foods.

Infants and food allergy: Dr. F. L. Grusky and his colleagues have found that premature newborn infants absorb much larger quantities of ingested food antigens than do older infants or children. [21] This suggests that mothers who are breast-feeding or starting to feed solid foods to their infants should be very cautious in exposing them to potentially allergy-producing food substances because of the greater permeability of their intestinal tract and the potential of foreign proteins being transferred into the blood and initiating allergy.

Not only does protein present a major problem with regard to digestion and assimilation, fat is also very difficult to digest. The fat-splitting pancreatic enzyme called lipase is the hardest to replace by the use of a pancreatin oral tablet because this enzyme is the most easily broken down in the stomach. The thorough digestion of fat is very important, not only for the prevention of steatorrhea, or fat in the stool, but also for the uptake of essential fatty acids, which is absolutely required for good health.

Essential fatty acids such as linoleic, linolenic, and arachidonic acid, are necessary for the manufacture of a class of hormones called prostaglandins. Prostaglandins are used for promotion of wound healing, immune defense, fertility and control of functions in virtually every cell in your body. In the absence of adequate essential fatty acid absorption, there may be a disruption in the prostaglandin levels and very severe symptoms may result. Conditions such as cystic fibrosis, which has been associated with fat malabsorption, also cause essential fatty acid deficiencies which produce degeneration of the pancreas and heart, kidney problems and immune function difficulties. Effective fat digestion and absorption for the cystic fibrotic patient are promoted by the addition of high-potency pancreatic digestive aids to the diet supplement.

Insufficient pancreatic output of these digestive enzymes, either in pancreatic disease or in chronic pancreatic insufficiency, can lead to such symptoms as vision abnormalities and changes in the retina. This appears to be a result not only of the malabsorption of essential fatty acids, but also the malabsorption of fat-soluble vitamins, in this case particularly vitamin A.

The gallbladder and absorption of the fat-soluble vitamins: The absorption of fat-soluble vitamins such as A, D and E requires both sufficient levels of the pancreatic enzyme lipase and the presence of emulsifying substances released into the small intestine by the gallbladder. The gallbladder bile is composed of lecithin, cholesteryl esters, cholesterol and bile salts, which are extremely important in the emulsification of fats so they can be taken up by the body. People with bile insufficiency are fat malabsorbers and are unable to absorb the essential fatty acids and fat-soluble vitamins appropriately, which can lead to an array of symptoms relating to deficiencies of these nutrients.

For effective bile to be manufactured, there have to be adequate levels of an amino acid called taurine, vitamin C, magnesium and copper. Deficiencies of any of these nutrients can lead to fat malabsorption. A more lengthy discussion of taurine and its nutritional functions can be found in the Health Guide in this series entitled *Octacosanol, Carnitine and Other "Accessory" Nutrients*.

It should now be clear that proper digestion and absorption of dietary protein and fat are extremely important, not only to provide adequate calories from the diet, but also to assure uptake of nutrients essential to such functions as general immune function, wound healing, and maintenance of cellular vitality. We know that incomplete protein digestion can lead to absorption of small protein fragments which may produce allergic responses, as was first discussed by Drs. Newey and Smyth in the 1950s, and it is also known that these small protein fragments when they enter into the bloodstream can produce effects on the biochemistry of the brain which may alter mood, mind, memory or behavior.[22]

Recently work done at the National Institutes of Health has shown that the incomplete protein breakdown of gluten from wheat and casein from milk may produce a class of substances very similar to the endorphins, analgesic and mood-modulating substances that are produced within the body. When these materials are produced by incomplete digestion of certain reactive proteins, they are termed exorphins—meaning produced outside the body itself—but still may be absorbed into the blood and produce very powerful mood-altering effects.[23] One probable example of this is the occasionally observed association of schizophrenia with gluten allergy.[24]

It is also clear that proper absorption of fat is absolutely essential for the uptake of essential fatty acids and the fat-soluble vitamins and that the absorption of fat is dependent upon both the proper secretion of pancreatic lipase and the adequacy of bile.

Intestinal phase of nutrient absorption: Once the needed enzymes have been secreted by the pancreas and bile has been added to the digesting food in the intestinal tract, releasing nutrients in usable form, it is up to the intestines to absorb these nutrients. As mentioned earlier, this is accomplished by the millions of fingerlike villi lining the wall of the small intestine, with different regions along the intestinal tract having specific affinities for different nutrients.

Dietary fiber is known to help improve the function and absorptive capacity of the intestinal tract; however, there is one component of unrefined vegetable materials which may prevent the absorption of some of the trace minerals: phytic acid, also called inositol phosphate. Phytic acid is found in unsprouted or nonyeast-risen breads and cereal grains and seeds. This material is used by the seed in the process of germination to make the high-energy phosphate bonds required for the new sprout. In the absence of sprouting, or yeast rising, this inositol phosphate or phytate remains in the food as a powerful binder of what trace elements it may contain.

Fiber content is highest in whole grains, especially the concentrated covering or exosperm of the grains and is moderately low in fruits and vegetables and very low in meats and refined food products. Because of this, many Americans are not getting adequate levels of fiber in their diets and are suffering from intestinal and absorptive disorders that are a result of the fiber insufficiency. Most investigators now suggest 20-25 grams of fiber (say three to five tablespoons of bran) daily for optimal intestinal function.

The vital intestinal phase of nutrient absorption is extremely important and seems to be aided by adequate levels of nutrients such as zinc, pantothenic acid and vitamin A in the diet. Irritable bowel symptoms—localized inflammation of the intestinal lining—which result in malabsorption syndrome may be a consequence of a deficiency of one or more of these nutrients, or a food-allergic inflammatory response.

Candida albicans and intestinal function: One intestinal infectious agent which is known to increase the inflammatory response of the intestinal tract is the yeast organism Candida albicans. This organism is present in all of us but generally does not flare up and create a crisis. When that does happen, the resulting infectious condition is called candidiasis or moniliasis. Dr. Orian Truss has recently indicated that chronic infections of Candida can be quite serious and produce a range of adverse symptoms which may impede the absorption of nutrients.[25]

The yeast organism Candida is dimorphic; that is, it can exist in two different forms. One is the noninvasive yeast-like form and the other is the invasive mycelioid form, characterized by "roots" which infiltrate the intestinal wall and make it more susceptible to inflammation and leakage of foreign proteins. This may account for why individuals who suffer from Candida infection have rampant food allergies.

There are a number of accepted medical treatments for chronic Candida infection, and it has recently been found that the nutrient biotin, when taken at the level of 200 micrograms three times a day, can be extremely useful in preventing the conversion of the yeast-like form into the very dangerous invasive mycelioid form.[26] Reinoculation of the intestinal tract with acid-producing bacteria such as Lactobacillus acidophilus and the use of biotin supplementation may be very helpful for the person who has Candida overgrowth and is suffering not only from intestinal disorders but from other symptoms, including brain biochemical changes, as a result of the increased load in the blood of foreign proteins absorbed from the intestinal tract.

Dr. Truss feels that many cases of depression, anxiety, and asthma-like symptoms may be caused by chronic Candida albicans infections, which may ultimately result in the individual being allergic to virtually everything. Such a patient may think that many foods have to be eliminated from the diet when really what needs to be done is restore intestinal membrane integrity by preventing the mycelioid form of Candida albicans from growing.

Effects of intestinal function on vitamin absorption: One vitamin, folic acid, is actually denatured by the use of orally administered pancreatin. Folic acid is made up of chemical linkages that resemble those found in protein and therefore the administration of digestive aids can actually cleave the folic acid molecule and render it biologically unuseful. Thus one should not administer a folic acid-containing supplement near the time when utilizing digestive aids, but rather at least three hours before or after.

It is also known that there is a variety of vitamins that are absorbed only when bound to specific carrier proteins in the intestinal tract. These binding proteins are produced by the intestinal cells and help promote the absorption of these nutrients. If the intestinal cells are inflamed or damaged, they may not be able to secrete adequate levels of the vitamin-binding factors, and this results in water-soluble vitamin deficiency. As was pointed out in the introduction, good nutrition is dependent not only upon putting adequate nutrients in your mouth, but also upon proper digestion and absorption.

Individuals who have inflamed intestinal mucosa may be unable to secrete the levels of carrier proteins necessary for absorption, and so the first approach toward management of these patients would be to improve intestinal function by removing any allergy-producing proteins or substances from the diet, increasing zinc, vitamin A and pantothenic acid, and barring hypersensitizing foods, such as very spicy or rich ones.

OPTIMIZING NUTRIENT DIGESTION AND ABSORPTION

We can use the information in this Guide to promote effective nutrient digestion, starting in the mouth, continuing with proper nutrient absorption throughout the whole length of the intestinal tract, and ending with the healthy elimination of waste products. Individuals who suffer from the maldigestion/malabsorption syndrome should consider the possibility of problems with stomach acid secretion in the gastric phase of digestion or pancreatic enzyme or bile secretion in the intestinal phase. Regulation of hormone secretion to stimulate acid production and pancreatic enzyme and bicarbonate output are extremely important, and it must be recalled that the state of mind can control the secretion of these hormones which ultimately may control digestion. An individual who is distressed psychologically may be stimulating through hormone secretion the production of stomach acid in an empty stomach, thus increasing the risk of ulceration.

Food has a tremendous ability to soak up acid and prevent the stomach from becoming excessively acidic. One should never administer acids on an empty stomach, and should try to prepare the stomach for the food by being relaxed and psychologically ready for the meal. Eating just because you think it's time and not when you are really ready to eat may have profound adverse impact upon the digestion and utilization of nutrients. Also, eating too much at too late an hour of the day may increase the load of undigested nutrients on the intestinal tract and set the stage for intestinal problems.

Lastly, it can be seen that the whole process of nutrient absorption is somewhat like a dog chasing its tail. If you are inappropriately nourished due to an intake of poor-quality foods, your stomach's acid-secreting cells may be unable to secrete enough acid, which causes poor absorption of nutrients, which makes you less nourished, which makes you less able to secrete stomach acid . . . which obviously results in less nutrient absorption. You start spiraling downhill to lower and lower levels of health, not knowing how to break the chain. The link breaker might be to improve the quality of the diet, to utilize stomach acid or pancreatic enzymes to promote better nutrient absorption, or a combination of both. This, in conjunction with the restoring of beneficial intestinal bacteria to the large intestine and increasing the relative amount of dietary fiber in the diet, can improve nutrient utilization and digestive function immensely.

Individuals who have for a long time been suffering from what appear to be the symptoms of malnutrition, even when consuming a presumably excellent diet, should look first to the problems of digestion and assimilation. These chronic problems of malutilization of nutrients may account for much of the chronic ill health that now plagues individuals who do not understand the important role that digestive enzymes and intestinal absorption play in establishing overall health.

It should be recalled that orally administered pancreatic enzymes are very powerful substances for the promotion of fat, protein and carbohydrate digestion but should not have to be relied upon for prolonged periods of time. If the individual's intestinal tract can be brought to optimal function, he should be weaned of the need for digestive aids, then maintained on a high-quality mixed diet low in fat, sugar and salt. The long-term dependency upon digestive aids may reduce the ability of the digestive system to do the work for itself, which is counter to the idea of encouraging it to regenerate itself. By constructing the diet to reduce excessively high salt and sugar content, reducing dietary fat and improving dietary fiber and the levels of such nutrients as vitamin C, vitamin A, zinc and pantothenic acid—which, though not discussed in this Guide, dealing as it does primarily with the role of enzymes, are indeed important in the digestive process—many people who now suffer from a chronic maldigestion/malabsorption syndrome may for the first time in years start to feel the full benefit of the nutrients in their diets.

FIGURE 1
The Digestive System

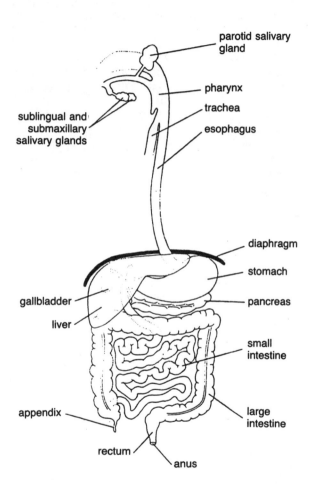

parotid salivary gland

pharynx

trachea

esophagus

sublingual and submaxillary salivary glands

diaphragm

stomach

gallbladder

pancreas

liver

small intestine

appendix

large intestine

rectum

anus

FIGURE 2
Potencies of Oral Pancreatic Enzyme Preparations

Preparation	Manufacturer	Type*	Enzyme Activity (μmol/unit)				
			Lipase	Trypsin	Chymo-Trypsin	Proteolytic Activity	Amylase
Ilozyme	Warren-Teed	T	3,600	3,444	2,807	6,640	329,600
Ku-Zyme HP	Kremers-Urban	C	2,330	3,082	2,458	6,090	594,048
Festal	Hoechst-Roussel	E	2,073	488	1,150	1,800	219,200
Cotazym	Organon	C	2,014	2,797	3,364	5,840	499,200
Viokase (Pan-5-Plus)	Viobin	T	1,636	1,828	2,081	4,440	277,333
Gastroenterase	Wallace	E	553	778	995	1,420	108,600
Ro-Bile	Rowell	E	539	661	735	980	68,000
Entozyme	A.H. Robins	E	495	668	483	1,088	39,000
Enzapan	Norgine	E	297	381	425	2,200	29,100
Phazyme	Reed & Carnrick	E	210	620	405	870	15,800
Ku-Zyme	Kremers-Urban	C	170	25	47	110	37,700
Digolase	Boyle	C	44	143	117	195	16,850
Arco-Lase	Arco	T	29	106	113	240	19,000
Convertin	BF Ascher	E	28	458	387	690	40
Kanulase	Dorsey	T	11	590	568	990	517
Zypan	Standard Process	T	10	296	313	410	441

*T represents tablet, C capsule, & E enteric-coated tablet.

from D. Graham, *N. Eng. J. Med.*, *296*, 1314 (1977).

REFERENCES

1. Lennard-Jones, J.E. 1983. Functional gastrointestinal disorders. *New Engl. J. Medicine 308,* 431.

2. Vander, A.J., Sherman, J.H. and Luciano, D.S. 1975. *Human Physiology,* second edition, New York: McGraw Hill.

3. Mellors, A. and Rose, R. 1977. Ascorbic acid flux across the human ileum. *Amer. J. Physiol. 233,* 374.

4. Tucker, D.M. and Inglett, G.E. 1981. Dietary fiber and personality factors as determinants of stool output. *Gastroenterol. 81,* 879.

5. Meyer, J.H. and Grossman, M.I. 1972. Comparison of D and L phenylalanine as pancreatic stimulants. *Amer. J. Physiol. 222,* 1058.

6. Meyer, J.H. and Way, L.W. 1970. Pancreatic bicarbonate response to various meals and acids in the duodenum. *Amer. J. Physiol. 219,* 964.

7. Mainz, D.L. and Webster, P.D. 1977. Effect of fasting on pancreatic enzymes. *Proc. Exp. Biol. Medicine 156,* 340.

8. Grossman, M.J. and Ivy, A.C. 1943. The effect of dietary composition on pancreatic enzymes. *Am. J. Physiol. 141,* 38.

9. Carlson, G.L. and Olsen, W.A. 1983. A bean-amylase inhibitor formulation (starch blocker) is ineffective in man. *Science 219,* 393.

10. Garrow, J.S. and Halliday, D. Starch blockers are ineffective in man. *Lancet,* January 18, 1983, p. 60.

11. Bo-Linn, G.W. and Fordtran, J.S. 1982. Starch blockers—their effect on calorie absorption. *New Engl. J. Medicine 307,* 1413.

12. Malagelada, J.R. and Go, V.L.W. 1979. Different gastric pancreatic and biliary responses to solid-liquid or homogenized meals. *Digestive Disease Science 24,* 101.

13. Hunt, J.N. and Oginski, A. 1962. The regulation of gastric emptying of various meals. *J. Physiol 163,* 34.

14. Hamilton, J.R. 1981. Retinal abnormalities with chronic pancreatitis. *New Engl. J. Medicine, 302,* 1316.

15. Graham, D. 1977. Comparison of exocrine pancreatic replacement enzyme activities. *New Engl. J. Medicine 296,* 1314.

16. DiMagno, E.P. 1977. Exocrine pancreatic replacement therapy. *New Engl. J. Medicine 296*, 1318.

17. Bray, G. 1941. Hypochlorhydria and childhood asthma. *Quart. J. Medicine 27*, 113.

18. Hemmings, W.A. and Williams, E.W. 1978. Transport of large breakdown products of dietary protein through the gut wall. *Gut 19*, 715.

19. Cornell, R. and Iselbacher, K.J. 1971. Small intestinal absorption of horseradish peroxidase. *Lab. Invest. 25*, 42.

20. Walker, W.A. 1974. Uptake and transport of macromolecules by the intestine—possible role in clinical disorders. *Gastroenter. 67*, 531.

21. Grusky, F.L., 1955. Gastrointestinal absorption of unaltered protein in normal infants. *Pediatrics 16*, 763.

22. Newey, H. and Smyth, D.H. 1959. Intestinal absorption of dipeptides. *J. Physiol. 145*, 48.

23. Wade, H. 1979. Exorphins from partial hydrolysis of gluten and casein. *J. Biological Chemistry 247*, 114.

24. Horrobin, D. Reconciliation of the dopamine, prostaglandin and opiod concepts in schizophrenia. *Lancet*, March 10, 1979, p. 59.

25. Truss, C.O. 1983. The missing diagnosis. Birmingham, Alabama.

26. Yamaguchi, H. and Iwata, K. 1982. Control of dimorphism in Candida albicans by differential biotin concentrations. *J. Bacteriol. 11-6*, 99.

- **Hypoglycemia** by Marilyn Light
- **Kelp, Dulse and Other Sea Supplements** by William H. Lee, R.Ph., Ph.D.
- **Lysine, Tryptophan and Other Amino Acids** by Robert Garrison, Jr., R.Ph., M.A.
- **The Mineral Transporters** by William H. Lee, R.Ph., Ph.D.
- **Nutrition and Exercise for the Over 50s** by Susan Smith Jones, Ph.D.
- **Nutrition and Stress** by Harold Rosenberg, M.D.
- **A Nutritional Guide for the Problem Drinker** by Ruth Guenther, Ph.D.
- **Nutritional Parenting** by Sara Sloan
- **Octacosanol, Carnitine and Other "Accessory" Nutrients Vol. 2** by Jeffrey Bland, Ph.D.
- **Propolis: Nature's Energizer** by Carlson Wade
- **Spirulina** by Jack Joseph Challem
- **A Stress Test for Children** by Jerome Vogel, M.D.
- **Tofu, Tempeh, Miso and Other Soyfoods** by Richard Leviton
- **Vitamin B3 (Niacin)** by Abram Hoffer, M.D., Ph.D.
- **Vitamin C Updated** by Jack Joseph Challem
- **Vitamin E Updated** by Len Mervyn, Ph.D.
- **The Vitamin Robbers** by Earl Mindell, R.Ph., Ph.D. and William H. Lee, R.Ph., Ph.D.
- **Wheat, Millet and Other Grains** by Beatrice Trum Hunter